Things Every Kid Should Know

REDUCE, REUSE & RECYCLE

ZAFAR NURI

EMAN
publishing

Eman Publishing
P.O. Box 404
FISHERS, IN 46038
www.emanpublishing.com

Order Online: www.ZafarNuri.com

ISBN 13: 978-1-935948-22-3
LCCN: 2011939276

Cover Design by Saqib Shaikh

Printed in the United States of America

Things Every Kid Should Know

REDUCE, REUSE & RECYCLE

ZAFAR NURI

Abu-Bakr and Ali are friends. They are both in the third grade. Ali spent all summer learning about 'reducing, reusing, and recycling' and Abu-Bakr didn't.

Ali learned about recycling at home from his parents. He learned that the average American uses about seven trees a year in things made, such as paper, pencils, and furniture.

He was amazed at how in America they use about 2,000,000,000 (2 billion) trees a year.

In Ali's house they have four garbage bins for recycling: 1 for glass, 1 for cans, 1 for plastic, and 1 for paper.

Ali learned that the most important thing that he and others needed to do first was to reduce how much they bought from the store.

"Too many things can make a clutter, waste lots of space, and make lots of garbage," said his mom.

One day when Abu-Bakr went over to Ali's house, they played, and they both drew on paper and cut out shapes. Ali told Abu-Bakr to reuse some of the paper to make airplanes.

Ali said, "When we reuse paper, we are able to help save some trees."

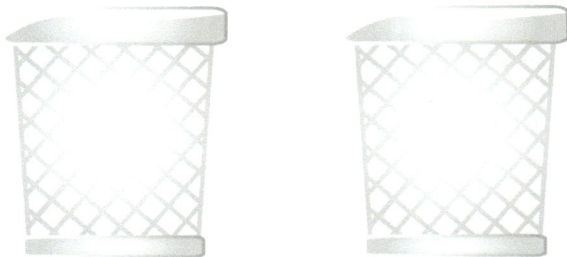

After they were done Abu-Bakr threw his extra papers in the trash can, and Ali threw his extra papers in the paper-recycling bin.

Ali then told Abu-Bakr, "Next time please put any type of used paper in the recycling bin. We have 3 recycling bins around the house just for paper. This is so that when we have used the paper, we just throw it in the recycling bin, instead of the trash can."

Abu-Bakr asked, "What is a recycling bin?"

Ali replied, "That is where we throw paper that we don't need. Then at the end of the week my father takes it to a recycling center and they recycle it all."

Abu-Bakr asked, "Why don't you throw it in the garbage can?"

Ali answered, "We recycle so that the paper can be used again."

"Wow, really? I did not know we can reuse paper again," said Abu-Bakr.

Then Ali told Abu-Bakr, "Did you know that if we all recycled our newspapers, that we could save almost 250,000,000 trees every year?"

Abu-Bakr told Ali that he wanted to recycle too. Ali smiled and told him that he needs a separate place for the items he wants to recycle.

From that day on Ali and Abu-Bakr would always recycle glass, paper, plastic, and cans.

Ali surprised Abu-Bakr and gave him a recycling ribbon for doing a good job at recycling.

What do the 3 R's mean?

a.   **Reduce**

b.   **Reuse**

c.   **Recycle**

1. **What does REDUCE mean?**
To use less.

2. **What does REUSE mean?**
To use something again for different purposes.

3. **What does RECYCLE mean?**
To use materials from certain items to make new items.

Why do we need to REDUCE?

Reduce means we don't waste anything; for example, buy enough vegetables to eat so that they don't rot, and so you don't have to throw them away. This will help save money, and make less garbage.

Why do we need to REUSE?

When we reuse things, we make less garbage. For example, we can reuse paper by writing on both sides of the paper. When the paper is all used up, we can even make paper airplanes.

Why do we need to RECYCLE?

Recycle means we use the glass, plastic, and other materials, that we have used before, to create something new; this helps us to not make so much waste, and to save the environment.

What are some items that everyone can recycle?

Cans

ALUMINUM CANS

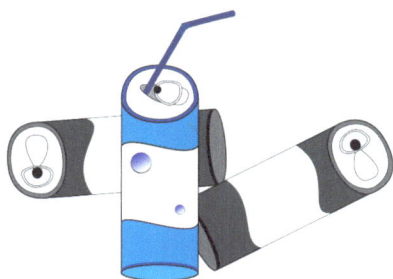

Paper

Glass

GLASS
BOTTLES

Plastic Bags

RECYCLE

PLASTIC
BAGS

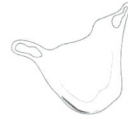

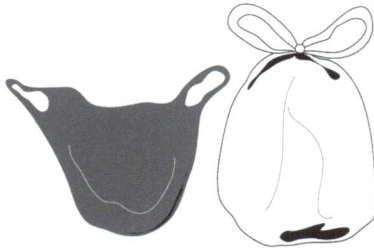

Batteries

BATTERIES

RECYCLE

Tires

Light Bulb (Save Energy)

Clothes

Kitchen Wastes

Eye Glasses

RECYCLE

Computer Wastes

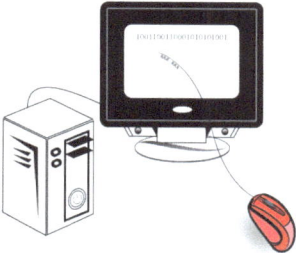

PDA and Cell Phones

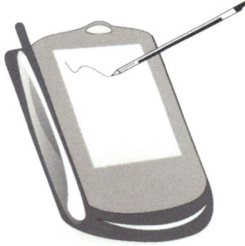

Electrical Stuff

ELECTRICAL

Grass

Books

Remember the list that I gave you so that you may Reduce, Reuse and Recycle. If we all do this together then WE can help save the Earth's resources!